# Maths Revision Booklet
## for CCEA GCSE 2-tier specification

## M2

*Lowry Johnston*

# Contents

A calculator may be used in these exercises.

Revision Exercise 1 ..................................................................... 3

Revision Exercise 2 ..................................................................... 10

Revision Exercise 3 ..................................................................... 17

Revision Exercise 4 ..................................................................... 24

Revision Exercise 5 ..................................................................... 31

Answers ..................................................................................... 38

# Revision Exercise 1

1. (a) Express as a power of 3:

    (i)  3×3×3

    Answer _____

    (ii) 9

    Answer _____

   (b) From the following list of numbers
   12, 24, 36, 72
   write down:

    (i)  the highest common factor (HCF)

    Answer _____

    (ii) the lowest common multiple (LCM)

    Answer _____

   (c) Jen saves her 1p and 2p coins. She has $x$ 1p coins in one jar and $x$ 2p coins in another jar. The total amount that she has in the two jars is £3.60
   Form an equation and use it to find the value of $x$.

   Answer $x =$ _____

2. A chef buys cooking oil which is delivered in cans which are cuboids. The cans measure 22 cm long, 15 cm wide and 31 cm high.

   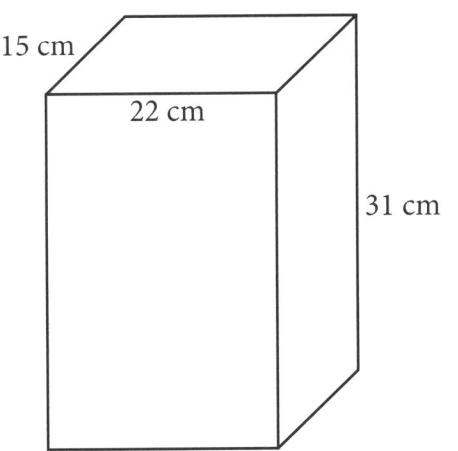

   (a) What is the volume of a can of cooking oil?

   Answer _____ cm$^3$

   (b) Write your answer correct to 2 significant figures.

   Answer _____ cm$^3$

3. **(a)** Hester pays £9.01 for a SIM only deal which allows unlimited texts and 500 minutes of phone calls per month. Her contract is for 24 months.
How much will she pay over the 24 months?

Answer £_____

**(b)** She decides to change her phone and is offered two options.
Option 1: continue paying £9.01 per month and buy a new phone for £110.00
Option 2: pay £15 per month for 24 months and get the phone free.

**(i)** Which is the best option over 24 months?

Answer _____

**(ii)** How much would she save by choosing this option?

Answer £ _____

4. **(a)** A taxi firm charges £2.50 per fare plus 50p per mile.

**(i)** How much would a 13 mile taxi fare from Bangor to Belfast cost?

Answer £ _____

**(ii)** How far did Mariam travel if her taxi fare cost £17.50?

Answer _____ miles

**(b)** The cost of tickets for a Pantomime are:

Adult £22.00      Child £11.00      Senior Citizen £15.00

A family spends £70.00 on tickets. They buy at least one of each type of ticket.
How many of each ticket do they buy?

Answer _____ Adult tickets

Answer _____ Child tickets

Answer _____ Senior Citizen tickets

5. (a) On a 10 day holiday a family spent:
   ⅖ of their time at the hotel pool
   30% of their time at the beach
   the rest of their time sightseeing

   How many days did they spend sightseeing?

   Answer _____ days

(b) The total cost of the family holiday was made up of accommodation, travel and extras as follows:
   Accommodation costs: 55% of the total cost
   Travel costs: ⅜ of the total cost
   Cost of extras: £330

   What was the total cost of the holiday?

   Answer £ _____

(c) They took 1500 euros spending money with them. If 1 euro equals £0.86, how much in £s did it cost (at 0% commission) to change their money into euros?

   Answer £ _____

6. Use the following flow diagram to sort the input numbers.

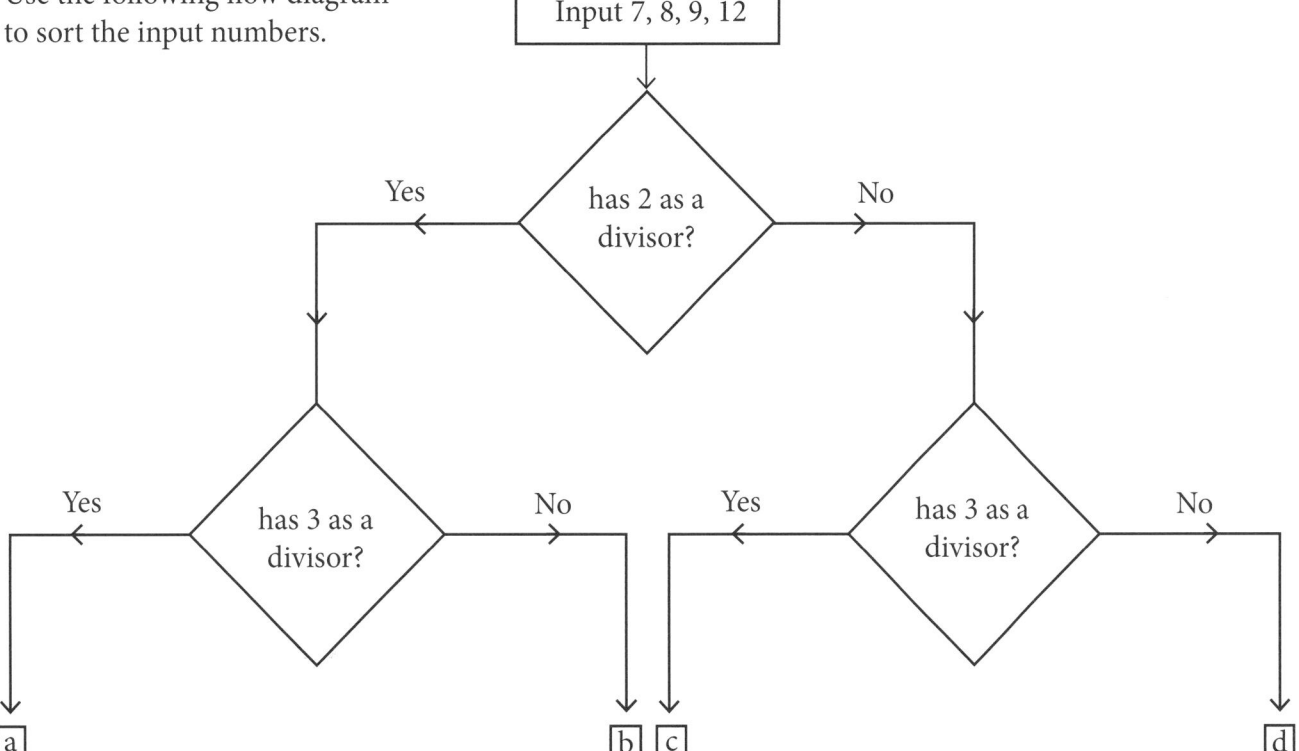

If the numbers 7, 8, 9 and 12 are input, which number is output at:

a = _____    b = _____    c = _____    d = _____

7. **(a)** Solve:
   
   **(i)** $\frac{x}{5} = 10$

   Answer $x = $ _____

   **(ii)** $2x + 5 = 10$

   Answer $x = $ _____

   **(b)** Expand:
   $5(x + 3)$

   Answer _____

8. **(a)** Write
   $0.9$, $\frac{19}{20}$, and $9\%$ in ascending order.
   **Show your working**.

   Answer _____ , _____ , _____

   **(b)** Orange squash is made from mixing 150 ml of orange with 450 ml of water.
   Write this as a ratio in its simplest form.

   Answer _____

   **(c)** A coffee drink is made up of milk, espresso coffee and cream.
   If ¾ of it is milk and ⅙ of it is espresso coffee, then what fraction of it is cream?

   Answer _____

**9.** A basin is made in the shape of a cuboid. Its length is 30 cm, its width is 30 cm and its height is 20 cm.

(a) How much water will it hold?

Answer _____ cm³

(b) When it is ¾ full what is the height of water in the basin?

Answer _____ cm

**10.** (a) Express as a power of 2:

(i) $2 \times 2 \times 2 \times 2 \times 2 \times 2$

Answer _____

(ii) 16

Answer _____

(b) A garden is made up as follows:
1/10 is flower beds
5/6 is lawn
the rest is paths

Calculate the fraction of the garden which is paths.
Give your answer in its simplest terms.

Answer _____

(c) In a survey 3 out of 8 people said they were good at maths.
Convert this to a percentage.

Answer _____ %

11. (a) Insert one of <, > or = to make each statement true.

   (i) $(0.1)^2$     0.001

   (ii) 20%     ⅕

   (iii) 0°C     −30°C

   (iv) ⅔     ¾

(b) James leaves Bangor at 13:50 and arrives in Belfast at 14:10
If the distance that he travelled was 13 miles, what was his average speed in mph?

Answer _____ mph

12. Because of major road works, Pat is unsure of how long it will take to get from home to work. The table lists the times taken, in minutes, during 3 working weeks (Monday to Friday).

| Day | Monday | Tuesday | Wednesday | Thursday | Friday |
|---|---|---|---|---|---|
| Week 1 | 21 | 27 | 31 | 38 | 29 |
| Week 2 | 23 | 19 | 32 | 27 | 41 |
| Week 3 | 32 | 28 | 32 | 37 | 19 |

(a) On how many days was the journey to work over 30 minutes?

Answer _____ days

(b) What time taken was the mode?

Answer _____

(c) What time was the median?

Answer _____

(d) What is the range of times for the journey to work?

Answer _____

(e) Describe what change you would expect in the median and range, if times were recorded for 3 weeks after the road works were completed.

Answer _____

13. Two Judges, X and Y, awarded marks to ten skaters in a competition. The marks are as follows.

| Judge X | 18 | 7 | 12 | 3 | 17 | 5 | 1 | 11 | 14 | 7 |
| Judge Y | 16 | 6 | 13 | 5 | 15 | 4 | 1 | 10 | 15 | 7 |

(a) Draw a scatter graph for these marks.

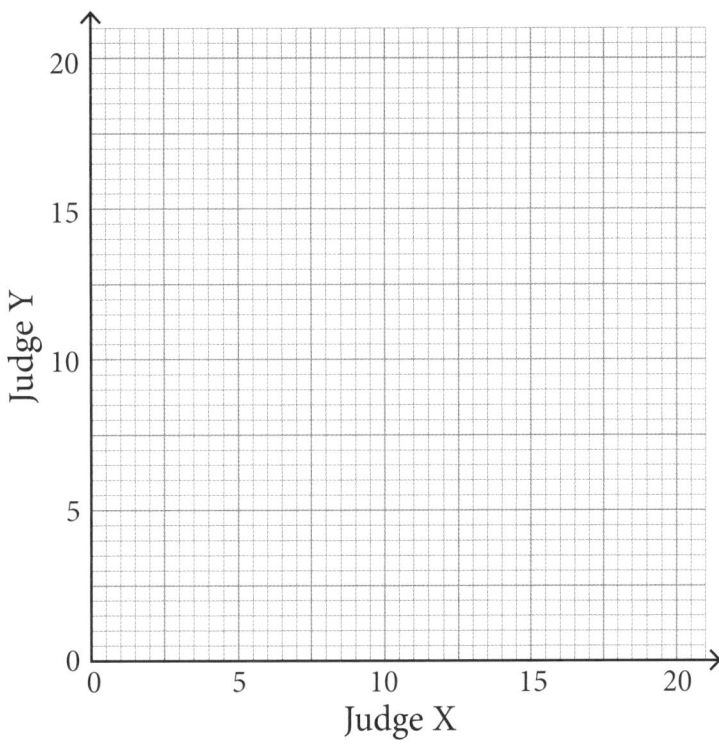

(b) Draw a line of best fit on the scatter graph.

(c) What type of correlation is represented by the line of best fit?

Answer _____

(d) A late entry was awarded 15 marks by Judge X. Estimate the mark that Judge Y may have awarded.

Answer _____

(e) Why might your estimate of the mark that Judge Y may have awarded be quite different from what he would actually mark?

Answer _____

# Revision Exercise 2

1. **(a)** In her will a mother leaves money to her three children. It is to be divided up as follows:
   Jim is to get 30% of it
   Mary is to get ⅓ of it
   John is to get £22 000

   **(i)** How much is Jim's share of the money?

   Answer £ _____

   **(ii)** How much is Mary's share of the money?

   Answer £ _____

   **(b)** Jean receives a quote of £640.00 for the renewal of her car insurance. She is given the offer of paying this by 10 equal direct debit payments. If she chooses to pay this way she will be charged an extra 7.5%. How much would her total monthly payment be?

   Answer £ _____

2. **(a)** A plumber charges a £25 call out fee plus £15 per hour he works.
   It takes him 2 hours to fit a tap and waste pipe which cost £39.99
   What is his total bill?

   Answer £ _____

   **(b)** The volume of a container is 168 m³. It is a cuboid.
   The length of the base is 7 m and the width of the base is 6 m.
   What is the height of this container?

   Answer _____ m

3. **(a)** Write 125 as a power of 5.

   Answer _____

   **(b)** Michael spends ⁵⁄₁₂ of his day working in the office. He decides to reduce the time he works by ⅙ of a day. What fraction of the day would he then be working? Give your answer in its simplest form.

   Answer _____

4.

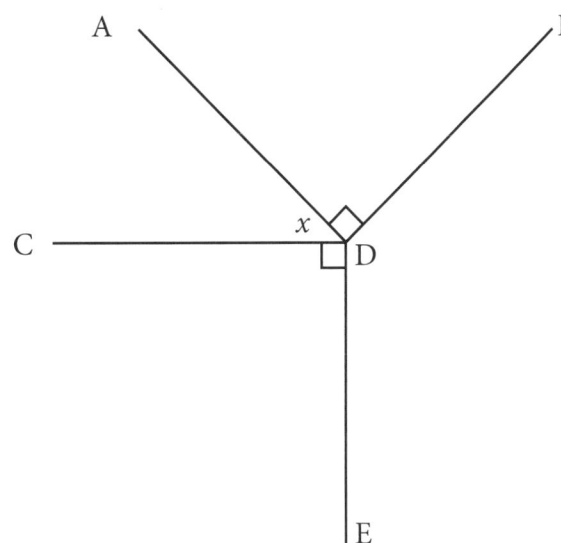

In the diagram above:
The angles ADB and CDE are right angles.
The angle BDE is twice the size of the angle ADC.

(a) Form an equation and use it to find the value of $x$.

Answer $x =$ _____

(b) A rectangle is $x$ cm long and 6 cm broad. Its area is $(2x + 32)$ cm². Form an equation and use it to find the value of $x$.

Answer _____

5. (a) Simplify:
   $4x + 3y + 2x - 5y$

Answer _____

(b) Solve:

(i) $\dfrac{x}{5} - 1 = 7$

Answer $x =$ _____

(ii) $2x + 9 = 1$

Answer $x =$ _____

6. (a) Plot the points A (2, 3) and B (5, 7) on the grid below.

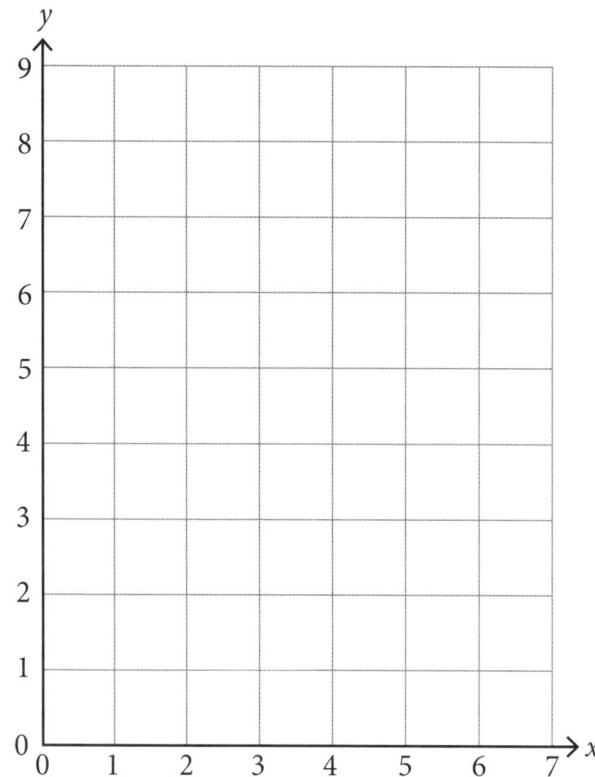

(b) Find the coordinates of the midpoint of AB.

Answer _____

(c) Find the length of AB.

Answer _____

7. (a) Members of a swimming club were asked what their favourite swimming event was.
The results are shown below.

| Backstroke | 7 |
| --- | --- |
| Breaststroke | 12 |
| Front crawl | 3 |
| Butterfly | 8 |
| Freestyle | 6 |

(i) What fraction, in its simplest terms, chose Butterfly as their favourite swimming event?

Answer _____

(ii) What percentage chose Breaststroke as their favourite event?

Answer _____ %

(b) Using the formula

Density = $\dfrac{\text{Mass}}{\text{Volume}}$

work out the density of iron, given that a 5 cm cube of iron weighs 975 g.

Answer _____ g/cm³

8. Use the following flow diagram to sort the input shapes.

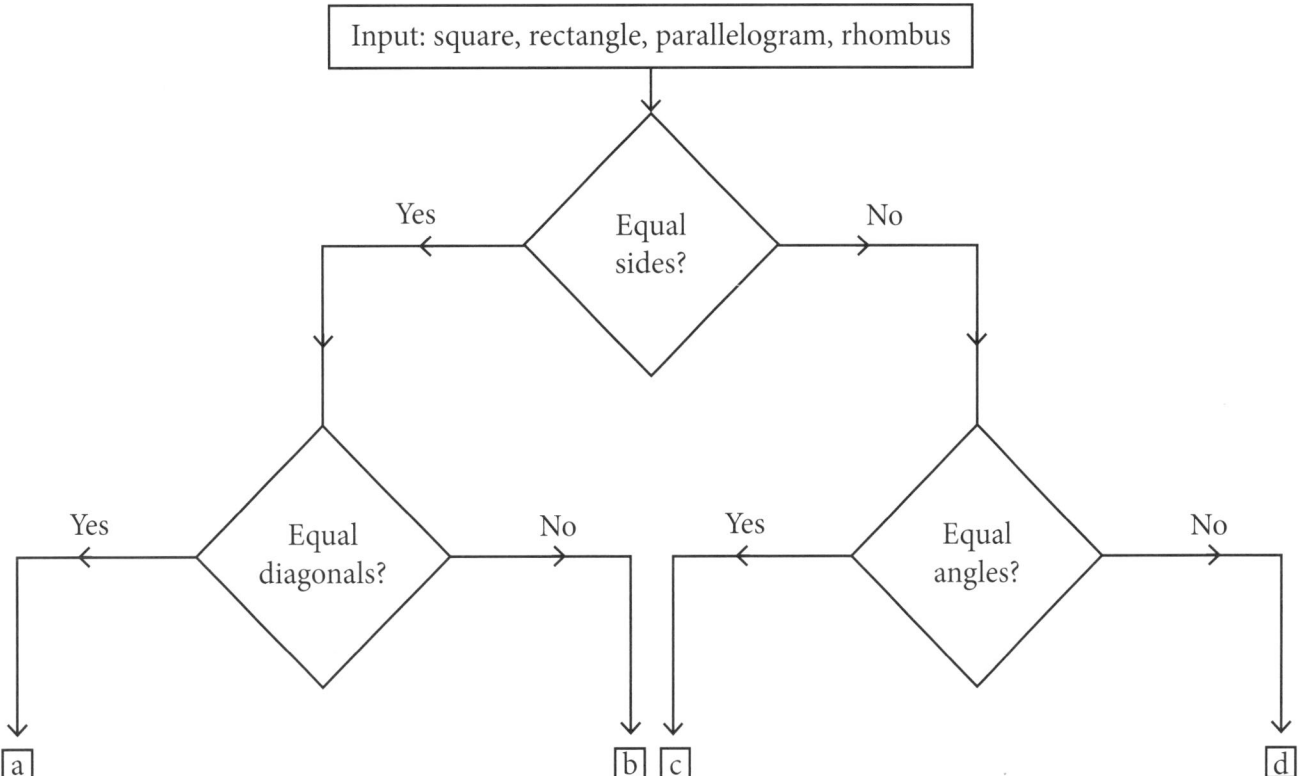

What shape will be output at:

a = _____

b = _____

c = _____

d = _____

9. Geography, History and French are subject options offered to 100 students.
   39 students chose Geography.
   46 students chose History.
   40 students chose French.

   Of these:
   12 chose Geography and History.
   9 chose Geography and French.
   10 chose History and French.
   3 chose Geography, History and French.

   (a) Complete the Venn diagram to represent the number of students choosing each subject.

   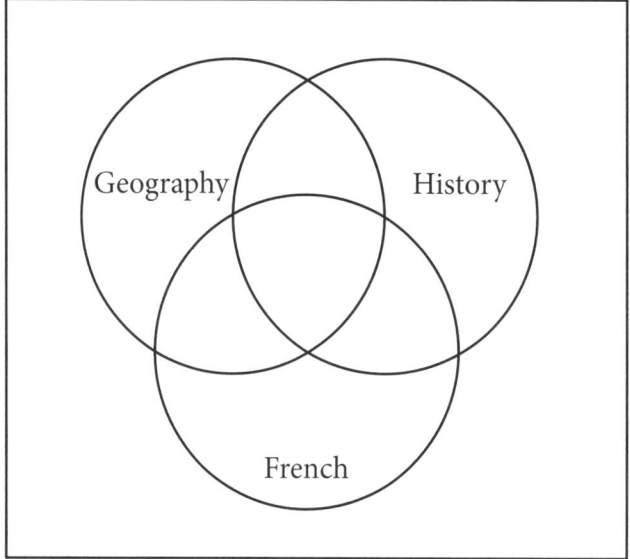

   (b) Calculate the number of students who have yet to make their choice.

   Answer _____

10. (a) The perimeter of a square side length $(x - 3)$ cm is equal to the perimeter of an equilateral triangle side length $(x + 1)$ cm.

    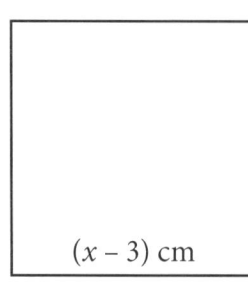
    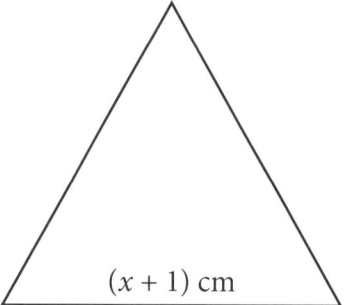

    Form an equation and use it to find the value of $x$.

    Answer _____ cm

(b) The Thompsons bought their house for £130 000 in 2016. Since then, house prices house prices have been increasing at 2% per year. What was the value of their house, 2 years later, in 2018?

Answer £ _____

11. (a) Plot the points A (−2, 3) and B (3, −9) on the grid below.

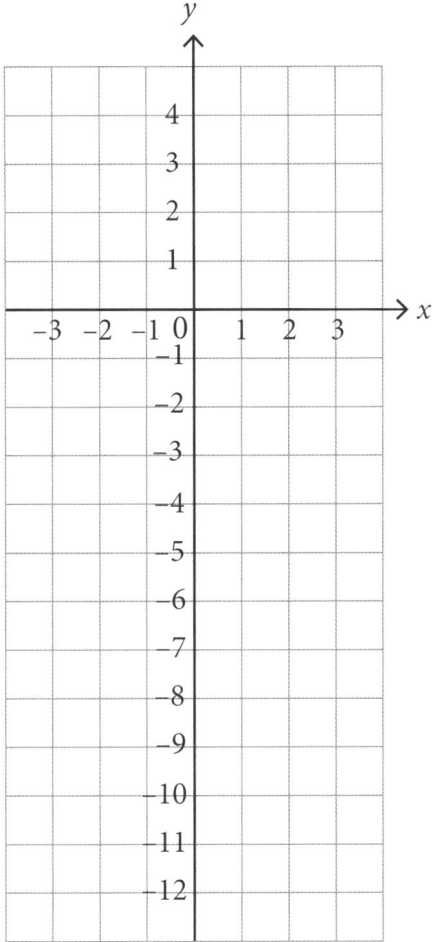

(b) Find the midpoint of AB.

Answer _____

(c) Calculate the length of AB.

Answer _____ units

12. (a) Calculate:

(i) $(0.1)^4$

Answer _____

(ii) $3^2 \times 10^3$

Answer _____

(b) Fruit is sold in cylindrical cans with the base area 38.5 cm² and height 10.5 cm.

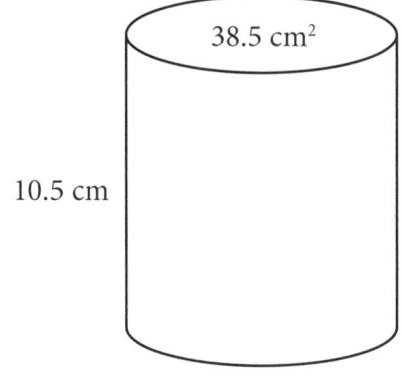

Use the formula:

Volume of a prism = area of cross section × length

to calculate the volume of the cylindrical can. Give your answer correct to 3 significant figures.

Answer _____ cm³

13. The diagram below shows the shape of pieces of metal cut out of 12 cm squares when manufacturing a toy. Find the area of metal cut out.

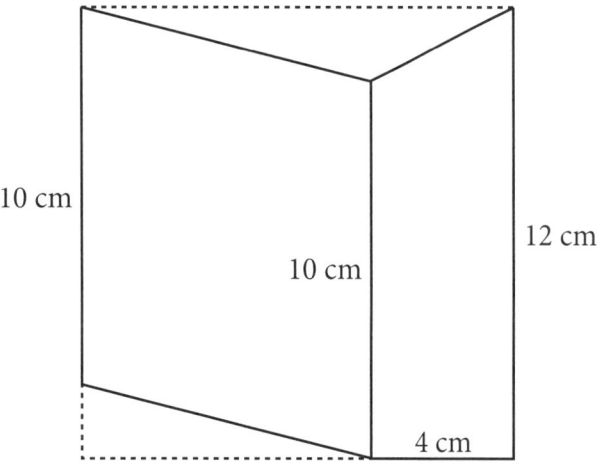

Answer _____ cm²

# Revision Exercise 3

1. **(a)** Faye runs 1.75 miles and then walks $\frac{3}{10}$ of a mile. Calculate the total distance that she travelled.

    Answer _____ miles

   **(b)** Jean uses $\frac{9}{10}$ of 500 ml of paint to paint a cupboard. Calculate how much paint she used.

    Answer _____ ml

   **(c)** Michael scored 180 out of 300 in a maths exam. Write this as a percentage.

    Answer _____ %

   **(d)** Calculate $2^3 \times 5^2$

    Answer _____

2. **(a)** Multiply out and simplify:
   $3(x - 5) + 1$

    Answer _____

   **(b)** Solve:
     **(i)**   $5(y - 3) = 20$

    Answer $y = $ _____

     **(ii)**   $2(3x - 1) = 28$

    Answer $x = $ _____

   **(c)** The perimeter of the rectangle shown below is 36 cm.

   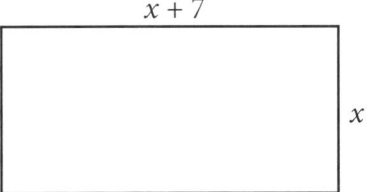

   Form an equation and use it to find the value of $x$.

    Answer $x = $ _____

3. Write the following recurring decimals as fractions in their lowest terms.

   (a) 0.5555…

   Answer _____

   (b) 0.2777…

   Answer _____

4. (a) ABCD is a parallelogram in which angle BAD = $x$ and ABC = $2x$.

   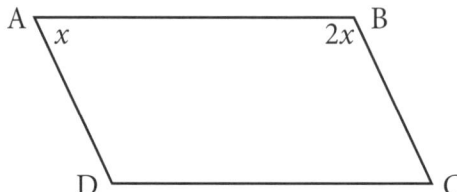

   Form an equation and use it to find the value of $x$.

   Answer _____

   (b) A pane of glass for a door is a trapezium ABCD in which angle BAD = angle ABC = 90°. The length of AB = 18 cm, the length of BC = 32 cm and the length of AD = 36 cm.

   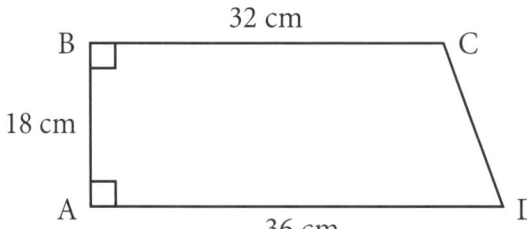

   Calculate:

   (i) the area of the pane of glass.

   Answer _____ cm²

   (ii) the length of DC correct to 3 significant figures.

   Answer _____ cm

5. Maria works in an office where she earns £16 800 per year.
   (a) How much is this per month?

   Answer _____

   She has to pay National Insurance at 12% on any money she earns above £8 400 per year.
   (b) How much National Insurance is deducted from her monthly pay?

   Answer _____

   She also has to pay Income Tax at 20% of any money she earns above £11 850 in a year.
   (c) How much Income Tax is deducted from her monthly pay?

   Answer _____

   (d) How much is Maria's monthly take-home pay?

   Answer _____

6. The pocket money given to a sample of children aged between 9 and 11 years old is shown below.

   | Pocket money (£P) | Frequency |
   |---|---|
   | $1 \leq P < 3$ | 1 |
   | $3 \leq P < 5$ | 14 |
   | $5 \leq P < 7$ | 27 |
   | $7 \leq P < 9$ | 39 |
   | $9 \leq P < 11$ | 6 |

   Calculate the mean pocket money.

   Answer £_____

7. In a year group of students there are 46 girls and 44 boys. A random sample of 15 students is selected. Their gender and marks in an English test and a Mathematics test, both marked out of 20, are recorded in the table below.

| Name | Gender | English | Maths |
|---|---|---|---|
| John | Male | 17 | 16 |
| Pat | Female | 11 | 10 |
| Rose | Female | 14 | 15 |
| Sean | Male | 10 | 18 |
| Ali | Female | 8 | 9 |
| Liam | Male | 15 | 15 |
| Jean | Female | 16 | 15 |
| Julie | Female | 19 | 20 |
| Anne | Female | 13 | 11 |
| Joe | Male | 17 | 10 |
| Bella | Female | 10 | 11 |
| Faye | Female | 12 | 13 |
| Dan | Male | 6 | 14 |
| Jill | Female | 18 | 19 |
| Callum | Male | 8 | 17 |

(a) Plot a scatter graph of the marks on the grid below.

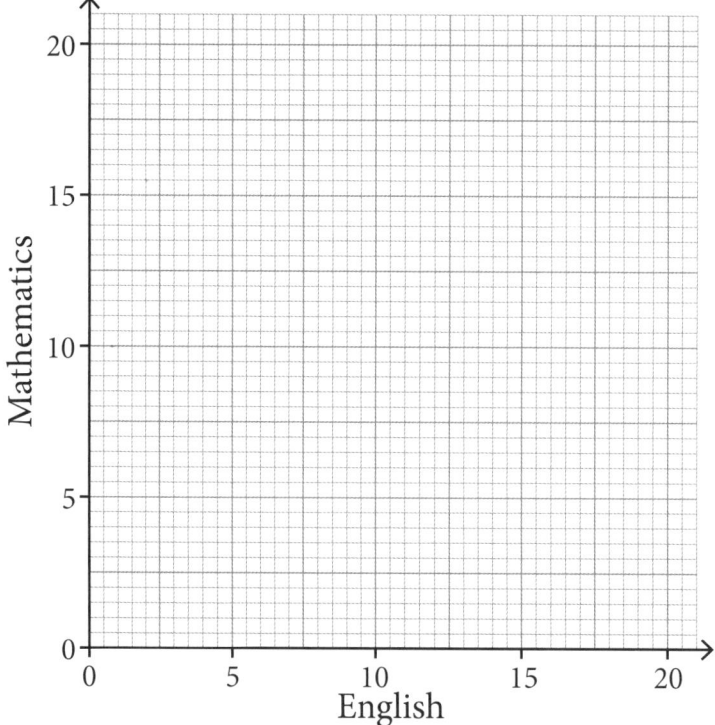

(b) How many outliers are there?

Answer _____

(c) Mark scored 18 out of 20 in the English test but missed the mathematics test. Should a line of best fit be used to find an estimate for his mark in Mathematics?

Answer _____

**(d)** Give a reason for your answer.

Answer _____

_____

_____

**(e)** Is the sample a true representation of the 90 students?

Answer _____

**(f)** Give a reason for your answer.

Answer _____

_____

_____

8. **(a)** Michelle bought a car for £16 000. In the first year it depreciated by 15%. In the second year it depreciated by 10% of its value at the end of year 1. What is the value of the car at the end of the 2 years?

Answer £ _____

**(b)** The Smith's town council rates bill for the year is £960. They can pay it by making 10 equal monthly payments by direct debit. There is a 2.5% charge for paying over the 10 months. How much is each monthly payment?

Answer £ _____

9. **(a)** In a vote for a new club secretary 70 out of 125 voted for Pat Brown. Write this as a percentage.

Answer _____ %

**(b)** Find the prime factors of 324. Then write 324 as a product of prime factors in index form.

Answer _____

**(c)** Tom cuts 2⅕ metres and 3¾ metres from a 15 metre reel of wire. Calculate how much wire is left on the reel.

Answer _____ metres

10. 100 people were asked which, if any, of three soaps (Emmerdale, Coronation Street and EastEnders) they liked to watch.
41 liked to watch Emmerdale.
50 liked to watch Coronation Street.
47 liked to watch EastEnders.

Of these:
9 liked to watch Emmerdale only.
30 liked to watch Emmerdale and Coronation Street.
26 liked to watch Coronation Street and EastEnders.
19 liked to watch all three soaps.

**(a)** Complete the Venn diagram to represent the number of people who liked to watch each soap.

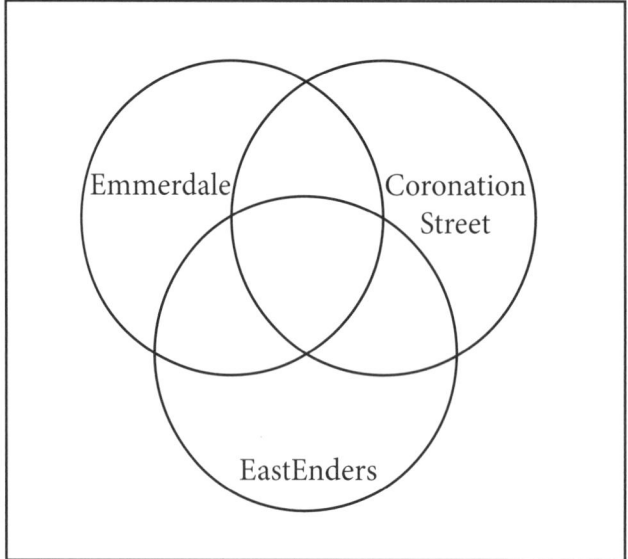

**(b)** Calculate how many people did not like to watch any of the soaps.

Answer _____

11. A kite maker has designed a kite, drawn below. By how much will the area of the kite change if the triangle at the top becomes equilateral by increasing the lengths of the bases of both triangles?

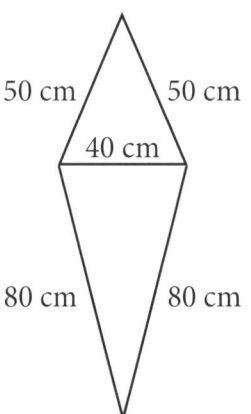

Answer _____

12. When phoning an information centre, callers may be put on hold depending on the volume of calls. For a sample of callers, the length of time on hold is recorded in the table below.

| Time ($t$) in minutes | $0 \leq t < 4$ | $4 \leq t < 8$ | $8 \leq t < 12$ | $12 \leq t < 16$ | $16 \leq t < 20$ | $20 \leq t < 24$ |
|---|---|---|---|---|---|---|
| Frequency | 32 | 20 | 14 | 2 | 3 | 1 |

(a) Which class contains the median?

Answer _____

(b) Which is the modal class?

Answer _____

(c) Calculate an estimate for the mean length of time a caller is put on hold. Give your answer correct to 3 significant figures.

Answer _____ minutes

13. (a) Expand and simplify:
   (i) $5(x + 2) + 2(x - 2)$

Answer _____

   (ii) $3(2x + 1) - 2(x - 1)$

Answer _____

(b) The areas of the two rectangles shown below are equal.

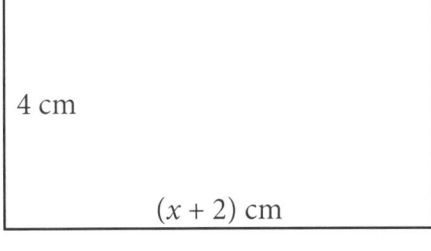

Form an equation and use it to find the value of $x$.

Answer _____

# Revision Exercise 4

1. **(a)** Multiply out and simplify:
   
   **(i)** $7 + 3(x - 2)$
   
   Answer _____
   
   **(ii)** $2(3x + 1) + 3(x - 2)$
   
   Answer _____

   **(b)** Solve:
   
   **(i)** $7(x + 3) = 35$
   
   Answer $x =$ _____
   
   **(ii)** $4(5x + 2) = 48$
   
   Answer $x =$ _____

2. John works in a factory where he earns £14 400 per year.
   
   **(a)** How much is this per month?
   
   Answer _____
   
   He has to pay National Insurance at 12% on any money he earns above £8 400 per year.
   
   **(b)** How much National Insurance is deducted from his monthly pay?
   
   Answer _____
   
   He has to pay Income Tax at 20% on any money he earns above £11 850 per year.
   
   **(c)** How much Income Tax is deducted from his monthly pay?
   
   Answer _____
   
   **(d)** How much is John's monthly take-home pay?
   
   Answer _____

3. The latitude and temperature at seven locations in the northern hemisphere are recorded in the table below.

| Latitude | 20 | 5 | 58 | 45 | 32 | 15 | 54 |
|---|---|---|---|---|---|---|---|
| Temperature (°C) | 35 | 42 | 20 | 28 | 31 | 37 | 22 |

(a) Draw a scatter graph for this data.

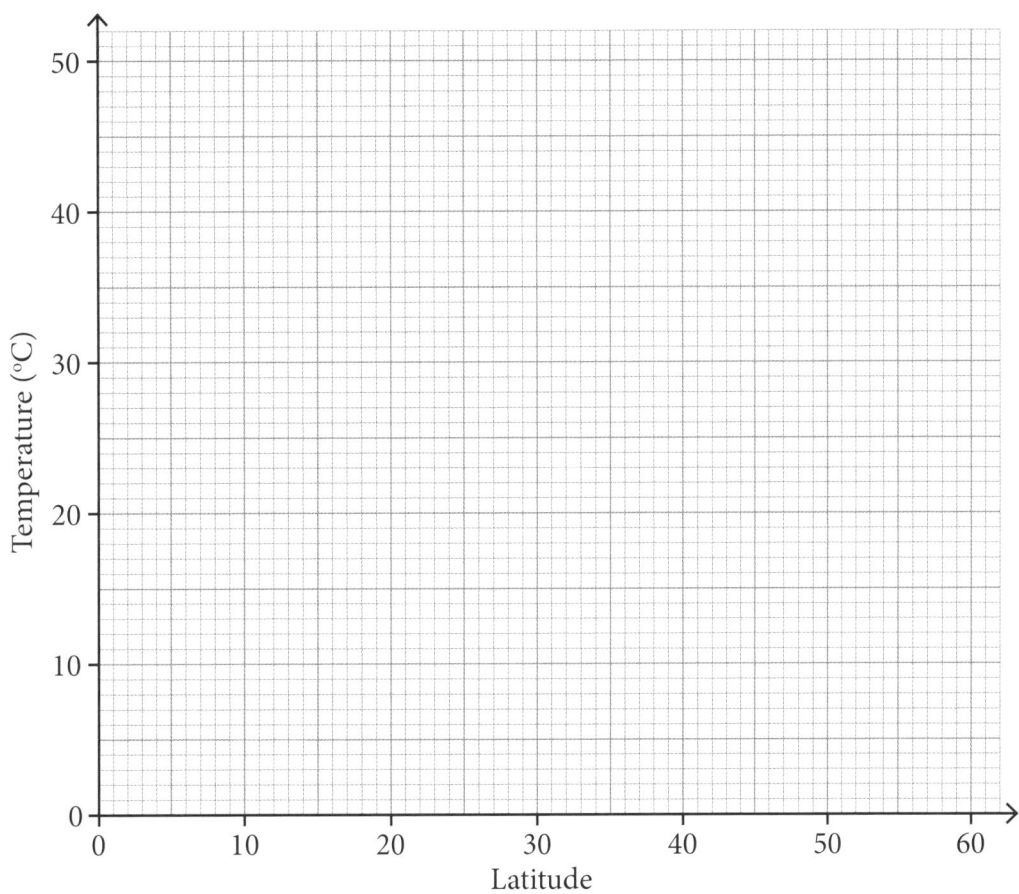

(b) Draw a line of best fit on the scatter graph.

(c) Use your line of best fit to estimate the temperature at the location where the latitude is 40°.

Answer _____

(d) What type of correlation does the graph show?

Answer _____

4. (a) Calculate the perimeter of the triangle shown below.

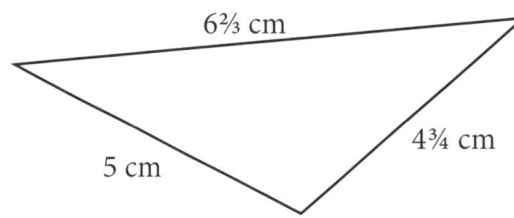

Answer _____ cm

(b) Calculate the compound interest on £300 invested for 2 years at 10% interest per year.

Answer  £ _____

5. (a) Look at the shape shown below.

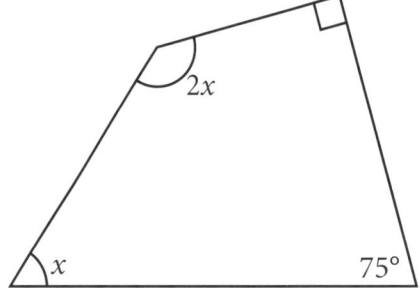

Form an equation and use it to find the value of $x$.

Answer  $x =$ _____ °

(b) (i) Find the perimeter of the triangle below. Write your answer in its simplest form.

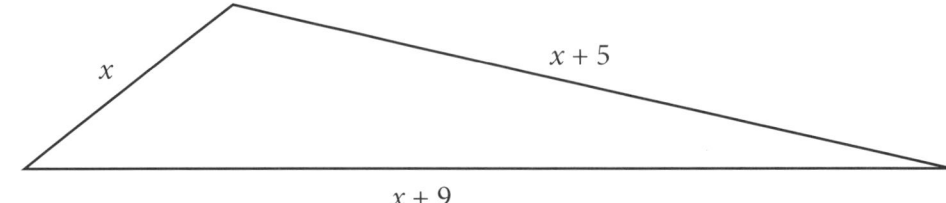

Answer _____

(ii) The perimeter of this triangle is 35 cm. Form an equation and find the value of $x$.

Answer $x =$ _____ cm

# Revision Exercise 4

6. **(a)** Write as a product of prime factors:
   **(i)** 30

   Answer _____

   **(ii)** 105

   Answer _____

   **(b)** What are the common prime factors?

   Answer _____

   **(c)** Find the HCF (highest common factor) of 30 and 105

   Answer _____

7. **(a)** In Eastwood School 380 pupils took canteen dinners each day. When healthy diet meals were started, the number of pupils taking canteen dinners increased by 40%. How many more pupils took canteen dinners each day?

   Answer _____

   **(b)** Mrs Jones wrote a cheque to pay Eastwood School £27.50 for her son's dinner tickets. Complete the cheque to show what she should have written.

   **NORDEN BANK**          98 – 76 – 54

   Norden plc, The Mall, Mainton, NB3 2AX       Date _____

   Pay _____

   _____   £ [        ]

                                          R. Jones

                                        _R Jones_

   ⑊000001⑊  98⎯7654⑊  12345678⑊

8. (a) Factorise:
   (i) $9x + 6$

   Answer _____

   (ii) $x^2 - 7x$

   Answer _____

   (iii) $x^2 + x$

   Answer _____

   (b) Solve:
   $\frac{x}{7} = 9$

   Answer $x =$ _____

9. (a) Which of the following fractions is nearest to $\frac{1}{5}$?

   $\frac{3}{20}$   $\frac{6}{25}$   $\frac{9}{40}$   $\frac{11}{50}$

   Show your working.

   Answer _____

   (b) When a tomato plant, 80 mm high, is given a fertiliser it grows by 12½% of its height in week 1.
   In week 2 it grows by 10% of its height at the end of week 1.
   What is the height of the tomato plant at the end of week 2?

   Answer _____ mm

   (c) Siobhan's dog eats ¾ tin of dog food per day.
   What is the least number of tins she should buy to last 1 week?
   Show your working.

   Answer _____ tins

# Revision Exercise 4

**10.** *Ampack* collects and delivers parcels. A random sample of 100 parcels is selected and their weights recorded. A summary is shown in the table below.

| Weight of parcel ($P$ kg) | Frequency |
|---|---|
| $0 \leq P < 5$ | 38 |
| $5 \leq P < 10$ | 25 |
| $10 \leq P < 15$ | 10 |
| $15 \leq P < 20$ | 15 |
| $20 \leq P < 25$ | 8 |
| $25 \leq P < 30$ | 4 |

(a) Which class contains the median?

Answer _____

(b) Calculate an estimate for the mean weight of the parcels.

Answer _____ kg

**11.** The graph below shows the amount, $A$ (£), in Sean's savings account for 5 consecutive weeks.

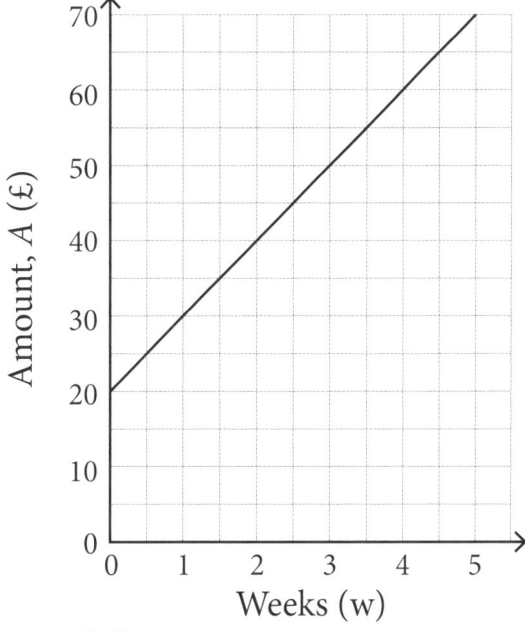

(a) Calculate the gradient of the straight line.

Answer _____

(b) Give a meaning to the value found in part (a).

Answer _____

(c) Explain the intercept in the Amount, $A$ (£) axis.

Answer _____

12. (a) Multiply out:
 $x(x^2 + 3)$

 Answer _____

 (b) Expand and simplify:
  (i) $5(2x + 3) - 2(3x - 5)$

 Answer _____

  (ii) $x(x - 2) + 4(x - 2)$

 Answer _____

 (c) Solve:
  (i) $7x - 8 = 3x$

 Answer $x = $ _____

  (ii) $8x - 13 = 3x + 7$

 Answer $x = $ _____

13. Membership of a Leisure Centre permits the use of the Gym, Swimming Pool and Squash Courts.
 100 members were asked which of these they used.
 46 used the Gym.
 58 used the Swimming Pool.
 34 used the Squash Courts.

 Of these:
 17 used the Gym and Swimming Pool.
 12 used the Gym and Squash Courts.
 32 used the Swimming Pool only.
 5 used the Gym, Swimming Pool and Squash Courts.

 Complete the Venn diagram to show the use each of Gym, Swimming Pool and Squash Courts.

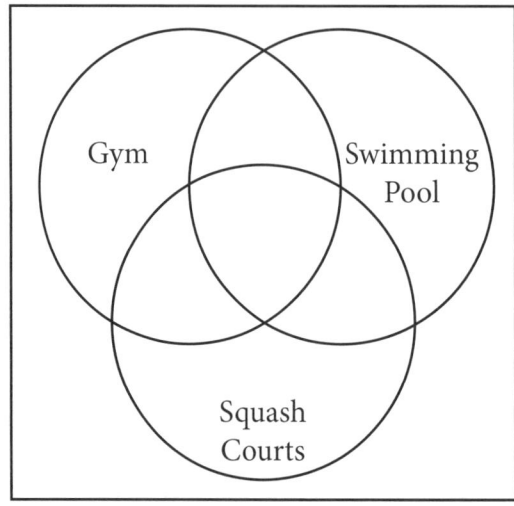

# Revision Exercise 5

1. (a) P and Q have coordinates (−3, 1) and (5, 5) respectively. Calculate:
   (i) the midpoint of PQ.

   Answer _____

   (ii) the length of the line PQ correct to 3 significant figures.

   Answer _____

   (b)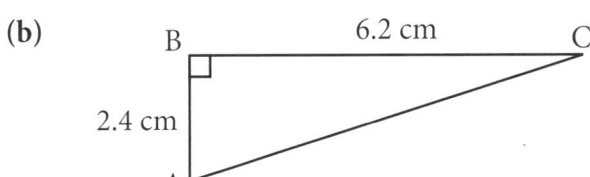

   ABC is a tile in a child's mosaic. It is a right angled triangle.
   AB = 2.4 cm and BC = 6.2 cm.

   Calculate:
   (i) the area of ABC,

   Answer _____ cm²

   (ii) the length of AC.

   Answer _____ cm

2. (a) Write the number 60 as a product of its prime factors.

   Answer _____

   (b) Find the LCM (lowest common multiple) of 60 and 42.

   Answer _____

   (c) Find the HCF (highest common factor) of 60 and 42.

   Answer _____

3. (a) Jenna and Neil have viewed a house priced at £150 000. If they intend to buy it, a deposit of 10% is required. To be able to get a mortgage for the amount that remains it must be less than 3 times their combined annual income.

   (i) How much is the deposit?

   Answer _____

   (ii) What is the least that their combined income must be to qualify for a mortgage?

   Answer _____

(b) They made an offer of £145 000 and it was accepted.
   (i) How much less would the deposit be?

   Answer _____

   (iii) How much less would their combined income need to be to qualify for a mortgage?

   Answer _____

4. Look at the shape below.

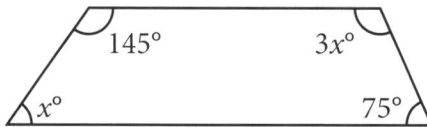

(a) Form an equation and use it to find the value of $x$.

Answer $x =$ _____

(b) Explain why this quadrilateral is a trapezium.

Answer _____

_____

_____

## Revision Exercise 5

**5.** The number of minutes each plane is late taking off from an airport is recorded for a sample of flights.

| Time (t) in minutes | Frequency | Mid-value |
|---|---|---|
| $1 \leq t < 9$ | 3 | 5 |
| $9 \leq t < 17$ | 17 | |
| $17 \leq t < 25$ | 28 | |
| $25 \leq t < 33$ | 31 | |
| $33 \leq t < 41$ | 16 | |
| $41 \leq t < 49$ | 9 | |

Calculate an estimate for the mean number of minutes that planes are late taking off.

Answer _____ minutes

**6.** Two shops have a sale on TVs.

**(a)**

*AB Electrical*

**SALE**

Flatscreen 32-inch TV

£660

We pay your 20% VAT

How much would this TV have cost if VAT had been included?

Answer £ _____

**(b)**

*RS Electrical*

**SALE**

Flatscreen TVs

15% off the marked price

The marked price of an identical 32-inch TV in *RS Electrical* is £750.
Which shop is the cheapest and by how much?

Answer Shop _____ by £ _____

7. *Plumbing and Co* charges a £40 call-out fee plus £20 per hour for the time spent on a job.
   *Pipe Works and Co* charges a £30 call-out fee plus £22 per hour for the time spent on a job.
   (a) How much would each charge for 2 hours work to fit a sink unit?

   (i) Plumbing and Co would charge £ _____

   (ii) Pipe Works and Co would charge £ _____

   (b) Both plumbing firms estimate that they would charge the same price for installing a bathroom suite.
   How many hours did they both plan to spend installing the bathroom suite?

   Answer _____ hours

8. (a) Ken invests £3600 for 3 years at 5% per year compound interest. What is the total amount after 3 years?

   Answer £ _____

   (b) In the circle drawn below, AC is a diameter and ABC is a right angle.

   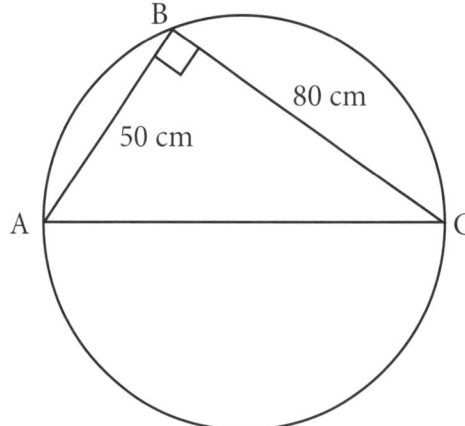

   Calculate the diameter of the circle.

   Answer _____ cm

9. A store buys dresses for £80.00 each. They are then priced so that the store makes a profit of 40% of this value on each dress sold. Customers with a store loyalty card are given 2% discount off the selling price. How much does a customer with a store loyalty card pay for **one** of these dresses?

Answer £_____

10. In a survey 239 out of 900 people said they liked *Marmite*. Express this as:
(a) A recurring decimal.

Answer _____

(b) A percentage correct to 2 significant figures.

Answer _____

11. The graph below shows the cost, $C$ (£), of hiring a car for $d$ days from *Fast Car Rentals*.

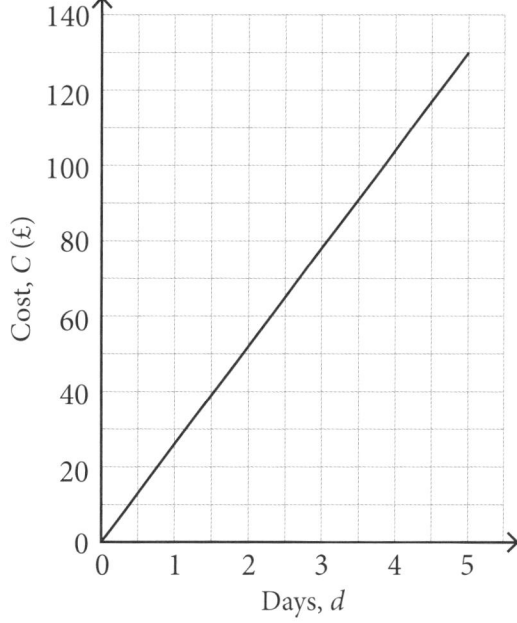

(a) Calculate the gradient of the straight line.

Answer _____

(b) Give a meaning to the value you found in part (a).

Answer _____

**12. (a)** The volume of a cylindrical paint tin is 5000 cm³.
How many litres is this?

Answer _____ litres

**(b)** If the height of the tin is 28 cm then what is the area of its base?
Give your answer correct to 3 significant figures.

Answer _____ cm²

**(c)**

| M Y Cash | Bank Account Statement | West Bank |
| 42 Uptown Road | | 40 High Street |
| Kinore | Your overdraft limit is £250 | Kinore |
| KN4 3PQ | | KN1 7GG |
| | Account Number 00120030 | |

| Date | Transaction Details | Payments Out | Payments In | Balance £ |
| --- | --- | --- | --- | --- |
| 8 February 19 | Balance brought forward | | | 87.22 |
| 9 February 19 | Bank Credit – Hut & Co. | | 309.62 | 396.84 |
| 10 February 19 | Direct Debit – Phone Co. | 65.98 | | 330.86 |
| 14 February 19 | ATM Cash Withdrawal | 200.00 | | _____ |
| 15 February 19 | Direct Debit – Electric Co | 74.35 | | _____ |
| 17 February 19 | Direct Debit – Car Loan | 109.00 | | _____ |
| 20 February 19 | Debit Card – Asha | 24.47 | | _____ |
| 21 February 19 | Lodgement | | 80.00 | 3.04 |
| 22 February 19 | Bank Credit – Hut & Co | | 309.62 | 312.66 |
| | Balance carried forward | | | 312.66 |

Complete the balance column of the bank statement to find between which two dates the balance in the account was below zero.

Answer Between _____ and _____

**13. (a)** The framework for the side of a shed is drawn below.
It is made from 1 slanted, 2 vertical and 3 horizontal pieces of wood.

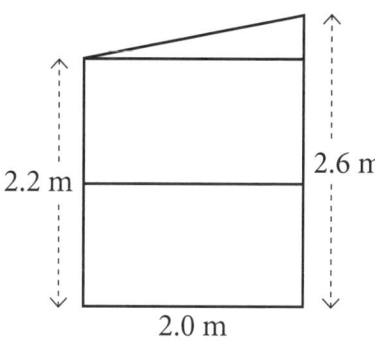

What is the total length of wood used?

Answer _____

**(b)** The wood is sold in 2.7 m lengths.
What is the percentage wastage? Give your answer correct to one decimal place.

Answer _____

# Answers

## Revision Exercise 1

1. (a) (i) $3^3$ (ii) $3^2$ (b) (i) 12 (ii) 72 (c) $3x = 360$, $x = 120$
2. (a) $22 \times 15 \times 31 = 10\,230$ (b) 10 000
3. (a) $9.01 \times 21 = £216.24$ (b) (i) Option 1 = 216.24 + 110 = 326.24; Option 2 = $15 \times 24 = 360$; so Option 1 is cheaper (ii) $360 - 326.24 = £33.76$
4. (a) (i) $2.50 + (13 \times 0.50) = £9.00$ (ii) $17.50 - 2.50$ fee = £15; £15 ÷ £0.50 per mile = 30 miles (b) One ticket each = $22 + 11 + 15 = £48$, leaving £22 more to spend; this is either 1 adult or 2 children; so: 1 adult, 3 child and 1 senior citizen tickets OR 2 adults, 1 child, 1 senior citizen tickets
5. (a) $10 \times 2/5 = 4$ days at pool; $10 \times 30/100 = 3$ days at beach; $10 - 4 - 3 = 3$ days sightseeing (b) Travel costs = $3/8 \times 100\% = 37.5\%$; accommodation and travel = $55 + 37.5 = 92.5\%$ of costs; so extras = 7.5% of costs; $330 ÷ 7.5 \times 100 = £4400$ (c) £1290
6. $a = 12$, $b = 8$, $c = 9$ and $d = 7$
7. (a) (i) $x = 50$ (ii) $x = 2.5$ (b) $5x + 15$
8. (a) $19/20 = 0.95$; $9\% ÷ 100\% = 0.09$; so correct order is 9%, 0.9, $19/20$ (b) 150:450 = 1:3 (c) $3/4 + 1/6 = 9/12 + 2/12 = 11/12$; cream = $1 - 11/12 = 1/12$
9. (a) $30 \times 30 \times 20 = 18\,000$ (b) $20 \times 3/4 = 15$
10. (a) (i) $2^6$ (ii) $2^4$ (b) $1/10 + 1/6 = 3/30 + 5/30 = 8/30 = 4/15$; paths = $1 - 4/15 = 11/15$ (c) $3/8 \times 100\% = 37.5\%$
11. (a) (i) > (ii) = (iii) > (iv) < (b) 14:10 − 13:50 = 20 minutes; 13 miles in 20 minutes = $13 \times 3 = 39$ miles in 60 minutes = 39 mph
12. (a) 7 days (b) 32 minutes (c) Arrange them in order; median = middle item = 29 minutes (d) $41 − 19 = 22$ minutes (e) Both to get smaller.
13. (a)

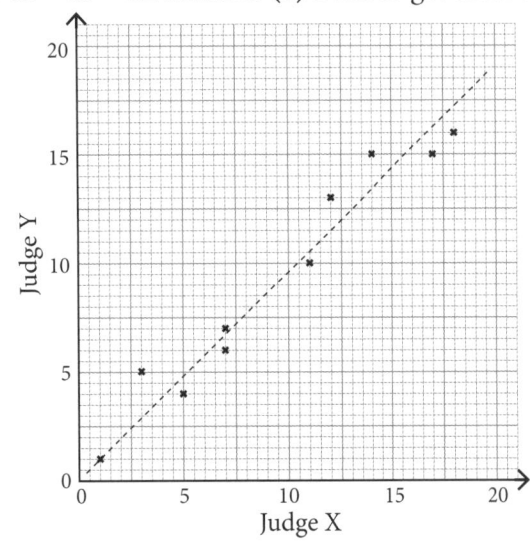

(b) See graph above – note line of best fit shown is approximate only (c) positive correlation (d) 14 (e) Judge Y need not be influenced by the mark Judge X awards.

## Revision Exercise 2

1. (a) (i) Jim = 30% = $3/10$; Jim + Mary = $3/10 + 1/3$ = $9/30 + 10/30 = 19/30$; So John = $1 − 19/30 = 11/30$; so total amount = $22\,000 ÷ 11/30 = £60\,000$; Jim's share = $60\,000 \times 30/100 = £18\,000$ (ii) $60\,000 \times 1/3$ = £20 000 (b) $640 \times 7.5/100 = 48$; so new cost = $640 + 48 = £688$; $688 ÷ 10$ months = £68.80
2. (a) $25 + 39.9 + (2 \times 15) = £94.99$
   (b) $168 ÷ (7 \times 6) = 4$
3. (a) $5^3$ (b) $5/12 − 1/6 = 5/12 − 2/12 = 3/12 = 1/4$
4. (a) $3x + 180 = 360$; $x = 60°$ (b) $6x = 2x + 32$; giving $x = 8$ cm (remember the units)
5. (a) $6x − 2y$ (b) (i) 40 (ii) −46. (a)

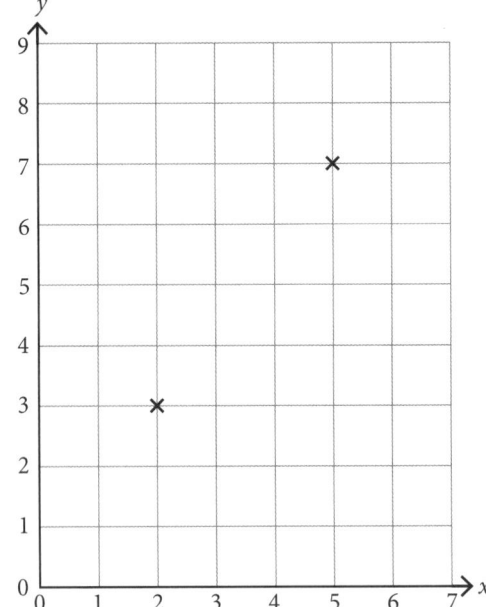

(b) (3.5, 5) (c) 5 (by Pythagoras' theorem)
7. (a) (i) $8/36 = 2/9$ (ii) $12/36 = 1/3 = 33\tfrac{1}{3}\%$
   (b) $975 ÷ (5 \times 5 \times 5) = 7.8$ g/cm$^3$
8. a = square, b = rhombus, c = rectangle, d = parallelogram

# Answers

9. (a)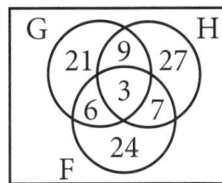

   (b) 3 students still to make their choice.
10. (a) $4(x - 3) = 3(x + 1)$, $x = 15$
    (b) $130\,000 \times 1.02 \times 1.02 = £135\,252$
11. (a)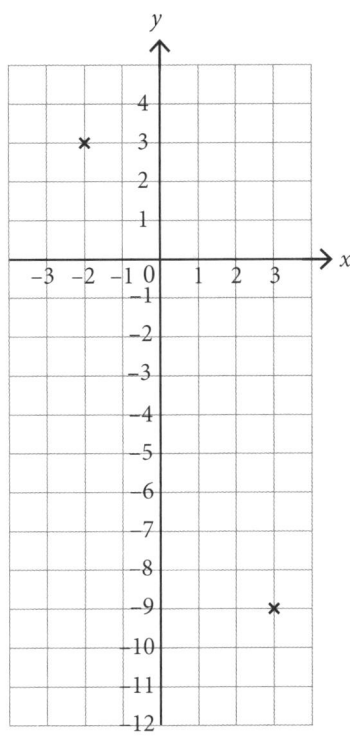

    (b) $(0.5, -3)$ (c) 13 (by Pythagoras' theorem)
12. (a) (i) 0.0001 (ii) 9000 (b) $38.5 \times 10.5 = 404$ cm$^3$
13. Area of square = $12 \times 12 = 144$; area of top triangle = ½ × base × perp height = ½ × 12 × 2 = 12; area of bottom triangle = ½ × 8 × 2 = 8; total removed = 12 + 8 = 20; so area cut out = 144 − 20 = 124 cm$^2$

## Revision Exercise 3

1. (a) $1.75 + 0.3 = 2.05$ (b) $500 \times 9/10 = 450$
   (c) $180 \div 300 \times 100\% = 60\%$ (d) 200
2. (a) $3x - 14$ (b) (i) 7 (ii) 5
   (c) $2(x + 7) + 2x = 36$; giving $x = 5.5$ cm
3. (a) Let $x = 0.555...$; so $10x = 5.555...$;
   $10x - x = 5.555... - 0.555...$; $9x = 5$; $x = 5/9$
   (b) Let $x = 0.2777...$; so $100x = 27.777...$; $100x - x$ = $27.777... - 0.2777...$; $99x = 27$; $18x = 5$; $x = 5/18$
4. (a) $x + 2x + x + 2x = 360°$, $x = 60°$
   (b) (i) Area of square = $18 \times 32 = 576$; area of triangle = ½ × 18 × 4 = 36; total area = 576 + 36 = 612
   (ii) Using Pythagoras' theorem: length$^2$ = $4^2 + 18^2$; giving length = 18.4
5. (a) $16\,800 \div 12 = £1400$ (b) $16\,800 - 8400 = 8400$; $8400 \times 12/100 = 990$; $990 \div 12 = £84.00$
   (c) $16\,800 - 11\,850 = 4950$; $4950 \times 20/100 = 990$; $990 \div 12 = £82.50$ (d) $1400 - 84 - 82.50 = £1233.50$
6. Total number of children = 87; total money assuming median value in each class = $(1 \times 2) + (14 \times 4) + (27 \times 6) + (39 \times 8) + (6 \times 10) = 592$; $592 \div 87 = £6.80$
7. (a)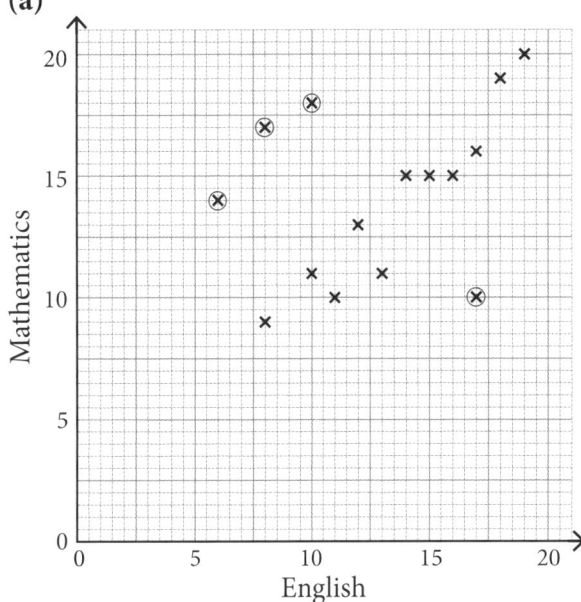

   (b) 4 (circled above) (c) No (d) A student who is good in English need not necessarily be good in Mathematics.
   (e) No (f) The ratio of girls to boys in the sample is 3:2 but in the population the numbers of girls and boys are almost the same.
8. (a) $16\,000 \times 85/100 = 13\,600$ after 1 year; $13600 \times 90/100 = £12\,240$ (b) $960 \times 2.5/100 = 24$; so total = $960 + 24 = 984 = £98.40$ per month
9. (a) $70 \div 125 \times 100\% = 56\%$
   (b) $2 \times 2 \times 3 \times 3 \times 3 \times 3 = 2^2 \times 3^4$ (c) $2\,1/5 + 3\,3/4$ = $2\,4/20 + 3\,15/20 = 5\,19/20$; $15 - 5\,19/20 = 9\,1/20$ or 9.05
10. (a)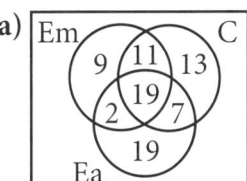

    (b) 20 people liked to watch none of the soaps.
11. Base of both triangles increases to 50 cm; perpendicular height of top triangle = 45.8 and bottom = 77.5 (by Pythagoras' theorem); area of original top triangle = ½ × 40 × 45.8 = 916; area of original bottom triangle = ½ × 40 × 77.5 = 1550; so total original area = 916 + 1550 = 2466; new area (by same method) = 3082.5; so area increases by 2982.5 − 2466 = 516.5 cm$^2$

12. (a) $4 \leq t < 8$ (b) $0 \leq t < 4$ (c) Total number of calls = 72; total length on hold assuming median value in each class = $(32 \times 2) + (20 \times 6) + (14 \times 10) + (2 \times 14) + (3 \times 18) + (1 \times 22) = 428$; $428 \div 72 = 5.94$
13. (a) (i) $7x + 6$ (ii) $4x + 5$
    (b) $4(x + 2) = 2(x + 5)$, $x = 1$

### Revision Exercise 4

1. (a) (i) $1 + 3x$ (ii) $9x - 4$ (b) (i) 2 (ii) 2
2. (a) £1200 (b) $14\,400 - 8400 = 6000$; $6000 \times {}^{12}/_{100}$ = £60.00 (c) $14\,400 - 11\,850 = 2550$; $2550 \times {}^{20}/_{100}$ = 510; $510 \div 12 = £42.50$ (d) $1200 - 60 - 42.50$ = £1097.50
3. (a)

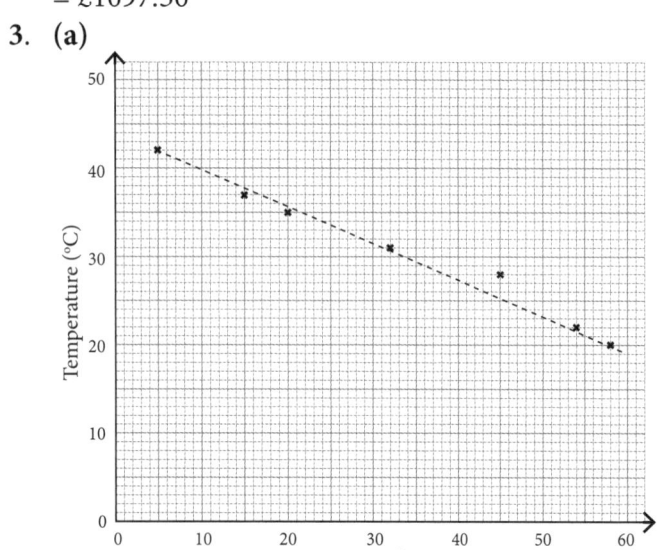

   (b) See graph above – note line of best fit shown is approximate only (c) 27° (depends on the line of best fit drawn) (d) negative correlation
4. (a) $5 + 6\frac{2}{3} + 4\frac{3}{4} = 5 + 6\frac{8}{12} + 4\frac{9}{12} = 16\frac{5}{12}$
    (b) $363 \times 1.10 \times 1.10 = £63.00$
5. (a) $3x = 360 - 75 - 90$; $x = 65°$ (b) (i) $3x + 14$
    (ii) $3x + 14 = 35$; $x = 7$
6. (a)(i) $2 \times 3 \times 5$ (ii) $3 \times 5 \times 7$ (b) 3, 5 (c) $3 \times 5 = 15$
7. (a) 152 (b) Words to read: Pay "Eastwood School twenty seven pounds and fifty pence", with "£27.50" written in the amount box and an appropriate date.
8. (a) (i) $3(3x + 2)$ (ii) $x(x - 7)$ (iii) $x(x + 1)$ (b) $x = 63$
9. (a) $\frac{1}{5} = 0.2$; $\frac{3}{10} = 0.15$; $\frac{6}{25} = 0.24$; $\frac{9}{40} = 0.225$; $\frac{11}{50} = 0.22$; so $\frac{11}{50}$ is closest to $\frac{1}{5}$ (b) $80 \times {}^{12.5}/_{100} = 10$; so height after 1 week = 90; $90 \times {}^{10}/_{100} = 9$; so height after 2 weeks = 99 mm (c) $\frac{3}{4} \times 7 = \frac{21}{4} = 5\frac{1}{4}$ tins, so 6 tins are needed
10. (a) $5 \leq P < 10$ (b) Total weight of all parcels assuming median value in each class = $(38 \times 2.5) + (25 \times 7.5) + (10 \times 12.5) + (15 \times 17.5) + (8 \times 22.5) + (4 \times 27.5) = 960$; $960 \div 100 = 9.6$ kg
11. (a) $(70 - 20) \div 5 = 10$ (b) The amount saved per week. (c) Sean already had £20 in his account at the beginning of the five week period.
12. (a) $x^3 + 3x$ (b) (i) $4x + 25$ (ii) $x^2 + 2x - 8$
    (c) (i) $x = 2$ (ii) $x = 4$
13.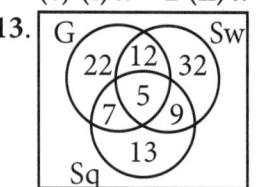

### Revision Exercise 5

1. (a) (i) (1, 3) (ii) 8.94 (by Pythagoras' theorem)
    (b) (i) $\frac{1}{2} \times 2.4 \times 6.2 = 7.44$ cm² (ii) 6.65 cm (by Pythagoras' theorem)
2. (a) $2^2 \times 3 \times 5$ (b) Prime factors of $42 = 2 \times 3 \times 7$; so LCM of 60 and $42 = 2 \times 2 \times 3 \times 5 \times 7 = 420$
    (c) HCF = $2 \times 3 = 6$
3. (a) (i) £15 000 (ii) Remainder = $150\,000 - 15\,000$ = £45 000 (b) (i) Deposit = £14 500; difference = $15\,000 - 14\,500 = £500$ (ii) Remainder = $145\,000 - 14\,500 = 130\,500$; $130\,500 \div 3$ = 43 500; difference = $45\,000 - 43\,500 = £1500$
4. (a) $x + 75 + 3x + 145 = 360$, $x = 35°$ (b) Both pairs of interior angles add up to 180°
5. Total number of planes = 104; total length of delays assuming median value in each class = $(3 \times 5) + (17 \times 13) + (28 \times 21) + (31 \times 29) + (16 \times 37) + (9 \times 45) = 2720$; $2720 \div 104 = 26.15$ minutes
6. (a) $660 \times {}^{20}/_{100} = 132$ VAT; $660 + 132 = £792$
    (b) In RS Electrical, $750 \times {}^{15}/_{100} = 112.50$; new price = $750 - 112.50 = 637.50$; so RS Electrical is cheaper by $660 - 637.50 = £22.50$
7. (a) (i) $40 + 2(20) = £80$ (ii) $30 + 2(22) = £74$
    (b) Let $x$ = number of hours; so $40 + 2x = 30 + 22x$; giving $x = 5$ hours
8. (a) $3600 \times 1.05 \times 1.05 \times 1.05 = £4167.45$ (b) 94.3 cm (or 94 cm to nearest cm) by Pythagoras' theorem
9. $80 \times {}^{40}/_{100} = £32$ profit; new price = $80 + 32 = 112$; 2% of this = $112 \times {}^{2}/_{100} = 2.24$; so price with 2% discount = $112 - 2.24 = £109.76$
10. (a) 0.26555… (b) $239 \div 900 \times 100\% = 27\%$ to 2 s.f.
11. (a) $130 \div 5 = 26$ (b) cost per day of hiring the car
12. (a) 1 litre = 1000 cm³, so $5000 - 1000 = 5$
    (b) Vol = height × base area; $5000 = 28 \times$ base area; giving base area = 179
    (c) 130.86, 56.51, –52.49, –79.96. Balance is below zero between 17 February and 21 February.